400+ COOL & UNBELIEVABLE ENGINEERING FACTS FOR KIDS

Contents

Introduction — 3
Chapter 1: Ancient Roman concrete — 5
Chapter 2: Self-healing materials — 10
Chapter 3: Biomimicry in engineering — 15
Chapter 4: The world's largest tunneling machines — 20
Chapter 5: Floating cities and sea-steading — 25
Chapter 6: Shape-memory alloys — 30
Chapter 7: Earthquake-resistant buildings — 35
Chapter 8: Solar-powered airplanes — 40
Chapter 9: Hyperloop transportation — 45
Chapter 10: Vertical farming technology — 50
Chapter 11: 3D-printed organs — 55
Chapter 12: Robotic exoskeletons — 60
Chapter 13: Space elevators — 65
Chapter 14: Invisibility cloaks and metamaterials — 70
Chapter 15: Maglev trains — 75
Chapter 16: Artificial intelligence in robotics — 79
Chapter 17: Bionics and prosthetic limbs — 83
Chapter 18: Underwater habitats — 88
Chapter 19: Smart dust and micro-sensors — 92
Chapter 20: Fusion reactors — 96
Chapter 21: Carbon capture technology — 101
Chapter 22: Swarm robotics — 105
Chapter 23: Nanomaterials and their applications — 110
Chapter 24: Tidal energy systems — 115
Conclusion — 120

Introduction

Hey there, young explorers and future inventors! Are you ready to dive into a world of mind-blowing science and incredible technology? Buckle up, because we're about to blast off on an adventure through the coolest engineering marvels you've ever imagined!

In this amazing book, ""400+ Cool & Unbelievable Engineering Facts for Kids,"" we're going to journey through 25 exciting topics that will make your jaw drop and your imagination soar. Have you ever wondered about ancient Roman concrete that's stronger than modern buildings? Or how about invisibility cloaks that could make you disappear like magic? We'll explore these and so much more!

Get ready to learn about robots that work together like ant colonies, buildings that can dance during earthquakes, and planes that fly using only the power of the sun. We'll dive deep into the ocean to discover underwater homes and zoom up to the stars in space elevators. You'll find out how scientists are printing human organs like they're pages in a book, and how tiny robots smaller than a grain of sand could change the world.

Imagine riding in a train that floats on air, or wearing clothes that can change color with your mood. We'll show you how engineers are creating artificial legs that can make people run faster than ever before, and how we might one day have cities that float on the ocean!

From capturing carbon to clean our air, to using the power of the tides to light up cities, this book is packed with incredible facts that will amaze your friends and family. You'll learn about cars that drive themselves, drones that can plant entire forests, and even how we might one day harness the power of the stars right here on Earth.

So, are you ready to become an expert in the coolest technology of today and tomorrow? Let's jump in and discover the unbelievable world of engineering together! Who knows, maybe you'll be inspired to become the inventor of the next big thing that changes the world!

Chapter 1: Ancient Roman concrete

1. Did you know that ancient Roman concrete is stronger than modern concrete? It's true! The Romans made their concrete with volcanic ash, lime, and seawater. This special mix has helped their buildings last for over 2,000 years! Some Roman structures are still standing today, like the Pantheon in Rome.

2. Imagine building a harbor that lasts for centuries! The Romans did just that with their special concrete. They poured it into wooden forms underwater, and it hardened into strong walls. These walls protected ships from big waves and storms. Today, scientists are studying this concrete to make our harbors better!

3. The secret to Roman concrete's strength? Seawater! When seawater mixes with the volcanic ash in the concrete, it forms tiny crystals. These crystals make the concrete super strong. It's like the concrete gets stronger over time instead of weaker!

4. Roman engineers were like super-smart chefs. They had a secret recipe for concrete that included volcanic ash, lime, and rock. This mixture was so good that it could even set underwater! That's how they built amazing harbors and sea walls that are still standing today.

5. Have you ever heard of self-healing concrete? Well, the Romans invented it first! Their concrete could fix tiny cracks all by itself. How? The mixture reacts with seawater to form new crystals that fill in the cracks. It's like the concrete has magic healing powers!

6. The Pantheon in Rome has the world's largest unreinforced concrete dome. It's been standing for almost 2,000 years! The dome is 142 feet wide and has a big hole in the top called an oculus. Rain and snow fall right through it, but the building is still strong. That's the power of Roman concrete!

7. Roman concrete was eco-friendly! It used less lime than modern concrete, which means it produced less carbon dioxide when made. The Romans didn't know about climate change, but their concrete was accidentally good for the environment. Maybe we can learn from their recipe to make greener buildings today!

8. Imagine concrete that gets stronger in seawater instead of weaker. That's Roman concrete! Scientists found that seawater creates new minerals in the concrete over time. These minerals, like Al-tobermorite, make the concrete super tough. It's like the concrete is exercising and getting stronger!

9. The Romans built a huge port called Caesarea Maritima in Israel using their special concrete. They made huge blocks underwater that weighed as much as 50 tons! These blocks are still there today, 2,000 years later. Modern divers can swim around them and see how amazing Roman engineering was.

10. Roman concrete had a secret ingredient: volcanic ash from Mount Vesuvius! This special ash helped the concrete set underwater and made it super strong. The same volcano that destroyed Pompeii also helped create some of the strongest buildings in history. Isn't that ironic?

11. Did you know that Roman concrete actually likes cracks? When tiny cracks form, seawater seeps in and starts a chemical reaction. This reaction forms new crystals that fill the cracks and make the concrete even stronger. It's like the concrete has its own repair kit!

12. Roman engineers were like ancient chemists. They mixed volcanic ash, lime, and rock in just the right way to make super-strong concrete. They even wrote books about their recipes! One famous Roman, named Vitruvius, wrote all about how to make the best concrete.

13. The Romans built huge fish tanks called piscinae with their special concrete. These tanks were right in the sea and held live fish for fancy Roman dinners. The concrete walls were so strong that many of these tanks still exist today, 2,000 years later!

14. Imagine building a road that lasts for centuries! The Romans did just that with their concrete. They used it to make the base of their famous roads, like the Appian Way. Parts of these roads are still used today, showing how durable Roman concrete really is.

15. Roman concrete was used to build amazing aqueducts that carried water for miles. These giant water bridges stood tall against earthquakes and storms. Some, like the Pont du Gard in France, are still standing today. You can even walk across them!

16. Scientists are like detectives trying to solve the mystery of Roman concrete. They use powerful microscopes and x-rays to look inside ancient Roman structures. They're discovering the secrets of this amazing material so we can make our buildings stronger and last longer.

17. The Romans used their special concrete to build underwater structures called piers. These piers extended out into the sea and helped create safe harbors for ships. Some of these piers are still visible today, showing how well Roman concrete works underwater.

18. Roman concrete was so good that it was used to build entire cities! The port city of Ostia, near Rome, was largely built with this amazing material. Even after being buried for centuries, many of the concrete buildings are still in great shape when archaeologists dig them up.

19. Did you know that Roman concrete is actually environmentally friendly? It uses less energy to make than modern concrete and lasts much longer. Some scientists think we should start using Roman-style concrete again to help fight climate change and build stronger buildings.

20. The Romans used their concrete to build massive domes, like the one on the Pantheon. This dome is still the world's largest unreinforced concrete dome! It's so well-designed that it has inspired architects for centuries. Even modern buildings copy its amazing structure.

Chapter 2: Self-healing materials

1. Imagine a bike helmet that can fix itself if it gets cracked! Scientists have created special plastics that can do just that. When the plastic breaks, tiny capsules inside it burst open. These capsules contain a liquid that fills the crack and hardens, making the helmet strong again. It's like having a tiny repair shop inside your helmet!

2. There's a new kind of concrete that can heal its own cracks! It has special bacteria mixed in that "sleep" until a crack appears. When water seeps into the crack, it wakes up the bacteria. The bacteria then make limestone, which fills the crack. It's like the concrete has its own team of tiny builders inside!

3. Did you know some paints can heal themselves? When these special paints get scratched, the scratch disappears on its own! The paint contains tiny capsules filled with extra paint. When scratched, these capsules break and release the paint, filling in the scratch. It's perfect for keeping cars and bikes looking shiny and new.

4. Scientists have created a self-healing rubber that works like our skin! When cut, this rubber oozes out a sticky liquid that glues the cut back together. After a while, the cut disappears completely. This could be great for making super-durable tires that rarely need replacing.

5. Imagine a phone screen that fixes its own cracks! Researchers are working on a special glass that can do this. When the screen cracks, a liquid flows into the crack and hardens, making the screen smooth again. No more worrying about dropping your phone!

6. There's a new type of fabric that can repair its own holes! It's made with special fibers that melt when heated. If you get a hole in your clothes, you just need to apply some heat, and the fibers will melt and close the hole. It's like having clothes that can sew themselves!

7. Scientists have made a metal that can heal cracks in itself! When the metal is damaged, you just need to heat it up. The heat makes the atoms in the metal move around and fill in the cracks. It's like the metal is playing a game of musical chairs to fix itself!

8. Imagine a spaceship that can fix holes made by space debris! Scientists are working on a special plastic for spaceships that releases a foam when punctured. This foam quickly hardens, sealing the hole and keeping the astronauts safe. It's like having a built-in repair crew in space!

9. There's a new kind of asphalt that can fix its own cracks! It contains tiny steel fibers and special particles. When you heat the road with a special machine, the fibers melt and mix with the particles to fill in cracks. It's like giving the road a warm hug to make it feel better!

10. Scientists have created a self-healing battery! When tiny cracks form inside the battery, a special liquid fills them up and hardens. This keeps the battery working well for longer. It's like giving the battery a superpower to fight against wear and tear!

11. Imagine a water pipe that can fix its own leaks! Researchers are developing pipes with a special coating inside. When a crack forms, this coating swells up and plugs the leak. It's like the pipe has its own little plumber living inside, ready to fix any problems!

12. There's a new kind of paint for airplanes that can heal small damages by itself! When the paint gets scratched, it releases a chemical that fills in the scratch. This helps protect the airplane

from rust and keeps it looking shiny. It's like giving the airplane its own healing superpower!

13. Scientists have made a self-healing electronic skin! This artificial skin can sense touch and temperature, just like our skin. If it gets cut, it can heal itself using special magnetic particles. This could help make better prosthetic limbs in the future. It's like creating a superhero skin!

14. Imagine a bridge that can fix its own cracks! Engineers are working on adding tiny capsules filled with healing agents to the concrete used in bridges. When a crack forms, these capsules break and release the agent, fixing the crack. It's like giving the bridge its own repair kit!

15. There's a special kind of plastic that can heal itself using light! When this plastic gets scratched, you just need to shine a special light on it. The light makes the plastic molecules connect again, closing up the scratch. It's like using a magic wand to fix things!

16. Scientists have created a self-healing coating for solar panels! This coating protects the panels from scratches and dust. If it gets damaged, the sun's heat makes it flow and cover the damage. It's like giving solar panels their own sunscreen that never wears off!

17. Imagine a house paint that cleans itself! This special paint breaks down dirt when sunlight hits it. Then, when it rains, the dirt washes away, leaving the house looking fresh and clean. It's like having a team of tiny housekeepers working on your walls all the time!

18. There's a new kind of glue that can unstick and restick itself! This glue is strong, but if you need to separate the things you've glued, you can. Then you can stick them back together again! It's inspired by geckos' feet, which can stick to walls and then unstick easily.

19. Scientists have made a self-healing plastic inspired by blood! When this plastic cracks, a red liquid flows into the crack, hardens, and fixes it. The liquid even turns white when it's done, like a healed scar. It's like the plastic has its own blood and healing system!

20. Imagine a car windshield that can fix its own chips and cracks! Researchers are working on a special layer for windshields that melts and flows into damage when heated. This could make driving safer and save money on windshield repairs. It's like giving your car the power to heal itself!

Chapter 3: Biomimicry in engineering

1. Did you know that the nose of Japan's bullet train was inspired by a kingfisher's beak? Engineers noticed how quietly kingfishers dive into water and thought, "What if we made trains like that?" They designed the train's nose to look like the bird's beak, and it made the train quieter and faster! Now that's a smart way to copy nature!

2. Imagine swimsuits that help you swim faster, just like shark skin! Scientists studied shark skin and found tiny V-shaped ridges that reduce drag in water. They created swimsuits with similar patterns, and swimmers wearing them broke world records! It's like wearing a shark costume that actually works!

3. Have you ever seen how easily a gecko can climb walls? Engineers were amazed by this and studied gecko feet closely. They found tiny hairs that help geckos stick to surfaces. Now, they're making super-strong adhesives inspired by gecko feet. Imagine being able to stick things together without any glue!

4. Nature has the coolest air conditioning system - termite mounds! These insects build tall mounds with clever tunnels that keep the inside cool, even in hot deserts. Architects in Zimbabwe copied this idea to design a shopping center that stays cool

without using much electricity. It's like having termites as your building designers!

5. Butterflies have beautiful, colorful wings, but did you know they're inspiring new technology? The way light bounces off butterfly wings creates those amazing colors. Scientists are copying this to make brighter screens for phones and computers, without using harmful chemicals. It's like turning your gadgets into butterfly wings!

6. Lotus leaves always stay clean, even in muddy ponds. How? Their leaves have tiny bumps that make water droplets roll off, taking dirt with them. Engineers copied this to make self-cleaning paints and fabrics. Imagine never having to wash your clothes again because they clean themselves just like lotus leaves!

7. Humpback whales have bumpy flippers that help them swim better. When scientists studied these bumps, they realized they could use the same idea to make wind turbines more efficient. Now, some wind turbines have whale-inspired blades that produce more electricity. It's like having a whale help power your home!

8. The boxfish looks kind of funny, but car designers think it's super cool! This fish's boxy shape actually helps it swim smoothly through water. Mercedes-Benz created a concept car shaped like

the boxfish, and it turned out to be very aerodynamic. Imagine driving a car that looks like a fish but cuts through the air like a knife!

9. Mosquito bites are annoying, but the way mosquitoes bite inspired something helpful! Engineers studied how mosquito mouths work and created a nearly painless needle for giving shots. The needle has tiny serrations like a mosquito's mouth, making it easier to pierce skin. It's turning a pesky bug into a medical hero!

10. Plants in the desert have clever ways to collect water. The Namibian desert beetle, for example, has bumps on its back that collect water from fog. Scientists copied this idea to make materials that can harvest water from the air, even in dry places. It's like having your own personal rainstorm wherever you go!

11. Woodpeckers can peck trees all day without getting headaches. How? They have special skulls that absorb shock. Engineers studied woodpecker skulls and used what they learned to design better helmets and protective gear. It's like having a woodpecker's superpower to protect your head!

12. Mussels can stick to rocks underwater, even in strong currents. Scientists studied the super-strong glue mussels make and are now creating waterproof adhesives that work even on wet

surfaces. Imagine being able to glue things together underwater, just like a mussel!

13. The kingfisher isn't just inspiring trains - it's helping solar panels too! The grooves on a kingfisher's beak inspired a design for solar panels that stay cleaner for longer. The special surface makes water droplets roll off, taking dust with them. It's like having a bird clean your solar panels for you!

14. Sharkskin isn't just inspiring swimsuits - it's fighting germs too! The pattern on shark skin makes it hard for bacteria to stick and grow. Scientists copied this pattern to make surfaces for hospitals that naturally resist germs. It's like having an army of tiny sharks protecting you from getting sick!

15. Owl wings are super quiet when they fly. Engineers studied owl feathers and found they have a special shape that reduces noise. Now, they're using this idea to make quieter wind turbines, fans, and even airplane wings. It's like giving everything owl superpowers to sneak around quietly!

16. The nose of a toco toucan inspired a cool invention for airplanes! This bird's beak is strong but light, thanks to a foamy bone structure inside. Engineers copied this to create new

materials for airplane parts that are strong but lightweight. It's like building planes with bird beaks!

17. Spider silk is super strong and stretchy. Scientists studied how spiders make their silk and are now creating similar materials in labs. These could be used to make super-strong, lightweight cables or even bulletproof vests. Imagine wearing a shirt as strong as a spider web!

18. Cats have tiny barbs on their tongues that help them stay clean. Engineers copied this idea to create a hairbrush that's gentler and more effective at detangling hair. It's like having a cat groom your hair - but without the cat spit!

19. The way prairie dogs build their burrows inspired a new way to cool buildings! Prairie dog tunnels are designed to create air flow that keeps their homes cool. Architects are now using similar designs to make buildings that stay cool naturally. It's like having prairie dogs as your air conditioning experts!

20. Fireflies inspired a new kind of LED light! The way firefly lanterns are shaped helps the light shine brighter. Scientists copied this design to make LED lights more efficient. Now, we can have lights that shine brighter while using less energy. It's like having a jar of fireflies light up your room!

Chapter 4: The world's largest tunneling machines

1. Meet "Bertha," the world's largest tunneling machine! She's as tall as a 5-story building and as long as a football field. Bertha dug a huge tunnel under Seattle, big enough for a highway. She's like a giant underground worm, munching through dirt and rocks to create paths for people to drive through.

2. Did you know that some tunnel boring machines have cool nicknames? There's "Big Becky" in Canada and "Martina" in Madrid. These massive machines are like underground superheroes, each with their own personality and special skills for digging through different types of earth.

3. Imagine a machine so big it has to be assembled underground! That's what happened with "Ada" and "Phyllis," two giant borers used to dig London's Crossrail tunnels. They were lowered into the ground piece by piece and put together like a giant puzzle. It's like building a spaceship, but underground!

4. The "Green Heart" tunnel boring machine in the Netherlands is special. It can dig through soft soil and hard rock! It's like having a spoon and a knife in one tool. This flexible machine helps engineers dig tunnels in tricky places where the ground keeps changing.

5. In China, a tunneling machine called "Ju Xing" set a world record. It dug 1,755 meters in just one month! That's like digging under 17 football fields in 30 days. This super-fast digger helps build subway systems quicker than ever before.

6. Some tunnel boring machines are like underground factories. They dig the tunnel, remove the dirt, and build the tunnel walls all at once! It's like having a cookie cutter that not only cuts the dough but also bakes and decorates the cookies at the same time.

7. The "Mega Mole" in Hong Kong can work underwater! It dug a tunnel connecting two islands. This machine is like a submarine and a digger combined. It can handle the pressure of water above it while still digging through the seafloor.

8. Did you know that some tunnel boring machines are so big they have kitchens and bathrooms inside? Workers can spend days inside these machines as they slowly dig through the earth. It's like living in a moving underground house!

9. In Japan, engineers built a tunneling machine that can turn corners! Most of these machines only go straight, but this one can curve as it digs. It's like having a car that can drive sideways to park in tight spots, but underground!

10. The "Tuen Mun - Chek Lap Kok Link" machine in Hong Kong is as heavy as 370 elephants! It dug a tunnel under the sea for cars to drive through. Imagine a machine so heavy it could squash hundreds of elephants, yet it floats on water as it digs!

11. Some tunnel boring machines have giant claws at the front called cutterheads. These claws spin around and break up rocks. It's like having a massive blender underground, turning solid rock into small pieces that can be easily removed.

12. In Seattle, engineers had to rescue "Bertha" when she got stuck underground. They dug a huge pit to reach her and fix her. It was like performing surgery on a giant robot! After two years of repairs, Bertha was ready to dig again.

13. The "Robbins XRE," used in China, can dig through hard rock and soft soil. It's like having a shovel that turns into a drill when it hits rocks. This flexible machine helps engineers dig tunnels in places where the ground keeps changing.

14. Some tunnel boring machines are so long that the crew at the front can't see the back! They use cameras and computers to coordinate their work. It's like driving a car so long that the driver needs a TV to see what's happening at the back bumper!

15. In Italy, a tunneling machine called "Marta" helped dig the longest railway tunnel in the world. It's so long that trains take 20 minutes to go through it! Imagine a tunnel so long you could watch a whole TV show while traveling through it.

16. The "Alaskan Way Viaduct Replacement Tunnel" machine in Seattle had to dig under buildings without disturbing them. It's like playing the game "Operation," but instead of a buzzer, you have to worry about a whole building falling down if you make a mistake!

17. Some tunnel boring machines use lasers to stay on course. These lasers help the machine dig in a straight line, even when it's deep underground. It's like having a super-accurate GPS that works underground to guide these giant diggers.

18. In Malaysia, a tunneling machine had to dig through a mountain full of caves! Engineers had to be extra careful not to disturb the caves or the animals living in them. It was like trying to dig a hole in Swiss cheese without breaking any of the holes!

19. The "Sissi" and "Heidi" machines in Switzerland dug the world's longest railway tunnel. They started digging from opposite ends and met in the middle! It's like digging a really long secret passage with your friend, starting from different houses and meeting in the middle.

20. Some tunnel boring machines can recycle the rock they dig up. They crush the rock and use it to make concrete for the tunnel walls. It's like if you could eat a cookie and then use the crumbs to make a new cookie right away!

Chapter 5: Floating cities and sea-steading

1. Imagine a city that floats on the ocean like a giant boat! Some engineers are designing these floating cities. They would have houses, schools, and parks, all bobbing gently on the waves. You could wake up every morning to the sound of seagulls and the smell of salty air!

2. In the Netherlands, there's a neighborhood of floating houses. When it rains a lot, the houses rise up with the water level. It's like living in a house that gives you a piggyback ride when it's wet outside! These clever homes stay dry no matter how high the water gets.

3. Scientists are thinking about building floating cities shaped like giant lily pads. These cities would have their own farms and use solar power. Imagine living on a huge leaf in the middle of the ocean, growing your own food and using the sun for electricity!

4. Some people want to build cities that can move around the ocean. These cities would have giant engines, like cruise ships. You could wake up near a tropical island one day, and a week later, be floating near icebergs! It's like your whole town going on a never-ending vacation.

5. In Japan, there's a plan for a floating city called Green Float. It would be as tall as Mount Fuji and home to 40,000 people! The city would float in warm tropical waters and have its own farm to grow food. It's like combining a skyscraper, a boat, and a farm all in one!

6. Oceanix City is a design for a floating town that can withstand hurricanes. The buildings would be low and spread out to stay balanced on the waves. Imagine going to school in a classroom that gently rocks back and forth, like being on a really big, comfy waterbed!

7. Some floating city designs include underwater tunnels connecting different parts of the city. You could ride in a glass tunnel and see fish swimming all around you on your way to the grocery store. It would be like living in a giant aquarium!

8. Engineers are creating floating gardens for these sea cities. Plants would grow on special rafts, their roots dangling in the seawater. Imagine picking fresh tomatoes from a garden floating right next to your floating house. It's like having a farm and a boat combined!

9. One idea for floating cities is to build them in a circle around a calm central lagoon. This lagoon would be perfect for swimming and water sports. It's like having a giant, protected swimming pool in the middle of your ocean neighborhood!

10. Some floating city designs include wave-powered generators. These machines would bob up and down with the waves to make electricity. It's like the ocean giving your city a never-ending supply of power just by moving around!

11. Architects have designed floating schools for these ocean cities. These schools would have big windows so students could watch sea life while they learn. Imagine doing your math homework while watching dolphins play right outside!

12. In some floating city plans, buildings would be made of light materials like bamboo. This would help the city stay afloat and be eco-friendly. It's like building a giant, comfy raft that thousands of people can live on!

13. One cool idea for floating cities is to have underwater hotels. You could sleep in a room with clear walls and watch colorful fish swim by all night. It would be like camping in the ocean, but cozy and dry!

14. Some designers want to create floating cities that clean the ocean as they move. They would have special systems to collect plastic and other trash from the water. Imagine living in a city that's like a giant vacuum cleaner for the sea!

15. In floating cities, people might use small electric boats instead of cars to get around. The 'roads' would be canals between buildings. It would be like living in a futuristic version of Venice, with everyone zipping around in eco-friendly water taxis!

16. One idea for floating cities is to have them change shape depending on the weather. In calm seas, the city would spread out. In storms, it would close up like a flower. It's like living in a city that can do yoga to stay safe!

17. Some floating city designs include special reefs underneath to attract sea life. You could look down through glass floors in your house and see colorful fish and coral. It's like having an aquarium under your feet wherever you go!

18. Engineers are working on ways to make food in floating cities using seawater. They could grow special plants that don't need freshwater and raise fish in floating ponds. Your whole dinner could come from the ocean around your floating home!

19. In some plans for floating cities, buildings would have sails to help the city move around. Imagine your whole neighborhood catching the wind like a sailboat, gliding across the ocean to visit new places!

20. One cool idea for floating cities is to have them clean and desalinate seawater for drinking. The city itself would act like a giant water filter. It's like living on top of your own water treatment plant, with fresh, clean water always available!

Chapter 6: Shape-memory alloys

1. Imagine a paper clip that can straighten itself out after you bend it! That's what shape-memory alloys can do. These special metals remember their original shape and can return to it when heated. It's like they have a built-in "undo" button!

2. There's a cool metal called Nitinol that can be used to make braces for teeth. When it's cold, dentists can easily bend it to fit your teeth. But once it warms up in your mouth, it slowly moves back to its original shape, gently pushing your teeth into the right position. It's like having a tiny robot straightening your teeth!

3. Some eyeglass frames are made with shape-memory alloys. If you sit on your glasses and bend them, you can dip them in hot water and they'll pop back to their original shape. It's like giving your glasses a warm bath to fix them!

4. Engineers use shape-memory alloys to make tiny robots that can squeeze through small spaces. When heated, these robots can change shape to move around obstacles. Imagine a robot that can slither through pipes like a snake to find and fix leaks!

5. In some cars, shape-memory alloys are used in air conditioning systems. They can control the flow of coolant without using electricity. It's like having a tiny, smart muscle in your car that knows when to flex to keep you cool!

6. Scientists have made artificial muscles using shape-memory alloys. These muscles can lift things much heavier than their own weight. Imagine having a robot helper with super-strong arms that can lift heavy objects easily!

7. Some spacecraft use shape-memory alloys to unfold solar panels in space. The panels are folded up for launch, but when they warm up in the sunlight, they unfold all by themselves. It's like having a giant space flower that blooms when it sees the sun!

8. There are special clothes hangers made with shape-memory alloys. You can fold them up small for travel, but when you hang them in a steamy bathroom, they pop back to their original shape. It's like having a magic hanger that grows when it gets warm!

9. Engineers have made heart stents using shape-memory alloys. These tiny tubes can be squished small to fit in blood vessels, but then expand to the right size when they warm up in the body. It's like having a tiny scaffold that builds itself inside your heart!

10. Some buildings in earthquake-prone areas use shape-memory alloys in their foundations. These metals can absorb the shaking and then return to their original shape, helping to protect the building. It's like giving a building its own shock absorbers!

11. There are toys made with shape-memory alloys that can move on their own when heated. You can bend them into funny shapes, then watch them wiggle back to normal in warm water. It's like having a toy that comes to life with a warm bath!

12. Scientists have created shape-memory fabrics that can change their texture when heated. Imagine a shirt that gets more breathable when you're hot, or a jacket that puffs up when it's cold. It's like wearing clothes that adapt to keep you comfortable!

13. Some fire sprinkler systems use shape-memory alloys. A special metal piece holds the water back until there's a fire. When it gets hot, the metal changes shape and lets the water out. It's like having a tiny firefighter waiting inside the sprinkler!

14. Engineers have made airplane wings that can change shape using these special metals. The wings can adjust to different flying conditions, making the plane more efficient. It's like giving an airplane the ability to flex its muscles mid-flight!

15. There are coffee makers that use shape-memory alloys to control water flow. When the water is hot enough, the metal changes shape and lets it flow over the coffee grounds. It's like having a tiny robot barista inside your coffee machine!

16. Scientists have created shape-memory alloy "muscles" for robots that work in space. These muscles can move without needing oil, which is important because oil would freeze in the cold of space. It's like giving space robots super-strong, freeze-proof arms!

17. Some shoe companies are experimenting with shape-memory alloys in running shoes. The metal could adjust the shoe's shape as you run, giving you the perfect fit all the time. It's like having shoes that mold to your feet with every step!

18. Engineers have used shape-memory alloys to make self-expanding screws for broken bones. These screws are small when inserted, but expand to the right size inside the body. It's like having a screw that grows to fix your bones perfectly!

19. There are shape-memory alloy engines that can run on small temperature differences. They could use the heat from your hand to power small devices. Imagine charging your phone just by holding it!

20. Scientists are working on shape-memory alloys that react to light instead of heat. These could be used to make smart windows that adjust their tint based on sunlight. It's like having windows that put on sunglasses when it's bright outside!

Chapter 7: Earthquake-resistant buildings

1. Did you know some buildings can dance during earthquakes? In Japan, there's a skyscraper that sits on giant balls! When the ground shakes, the balls roll a little, letting the building move without falling over. It's like the building is doing a slow dance to stay safe!

2. Imagine a building that floats on air during an earthquake! Some engineers have designed buildings that sit on special air cushions. When the ground shakes, the building gently floats above the moving earth. It's like giving a building its own hoverboard!

3. In California, there's a bridge that's held together with giant rubber bands! Well, not exactly rubber bands, but big rubber pads that act like shock absorbers. When an earthquake hits, these pads stretch and squish, keeping the bridge from breaking. It's like putting a big rubber eraser between the bridge and the shaking ground!

4. Some buildings in earthquake zones have special walls that can crumple like a car's bumper. These walls absorb the earthquake's energy, protecting the rest of the building. It's like giving a building its own set of airbags!

5. Engineers have created buildings with giant pendulums on top! When an earthquake hits, the pendulum swings in the opposite direction of the shaking, helping to balance the building. It's like having a huge Newton's Cradle toy keeping your building steady!

6. In Mexico City, there's a skyscraper that's like a building within a building! The inner building is connected to the outer one with flexible joints. During an earthquake, the two parts can move separately, reducing damage. It's like wearing a protective suit made of buildings!

7. Some buildings have special floors that can slide during earthquakes. These floors are connected with flexible joints and sit on smooth surfaces. When the ground shakes, they slide a bit, absorbing the movement. It's like giving a building roller skates to glide through earthquakes!

8. Imagine a building that can change its stiffness during an earthquake! Some smart buildings have computers that detect shaking and adjust the building's structure in real-time. It's like a building doing yoga to stay flexible during an earthquake!

9. In Japan, some houses are built on big springs! These springs compress and expand during earthquakes, reducing the shaking felt inside. It's like putting your whole house on a giant pogo stick to bounce through earthquakes!

10. Engineers have designed buildings with special "sacrificial" walls. These walls are designed to break in a controlled way during big quakes, absorbing energy and protecting the main structure. It's like giving a building its own set of crash test dummies!

11. Some skyscrapers have giant water tanks on top to help during earthquakes. The sloshing water helps counteract the building's swaying. It's like putting a giant waterbed on top of a building to keep it steady!

12. In California, there's a hospital that can move up to 3 feet in any direction during an earthquake! It sits on 150 special bearings that let it slide smoothly when the ground shakes. It's like putting the whole hospital on a giant air hockey table!

13. Imagine a building that can heal its own cracks after an earthquake! Some engineers are developing special concrete that can fix small cracks all by itself. It's like giving a building a superpower to heal its own wounds!

14. Some buildings have "seismic dampers" that work like shock absorbers in a car. They absorb the earthquake's energy, making the building shake less. It's like giving a building its own set of super-strong muscles to fight against earthquakes!

15. In Tokyo, there's a building shaped like a pagoda that's survived many earthquakes. Its secret? A central pillar that can sway independently from the rest of the building. It's like having a tree trunk that can bend in the wind while the branches stay still!

16. Engineers have created buildings with special "fuses" that break during big quakes. These fuses absorb energy and can be easily replaced after the earthquake. It's like giving a building a safety switch that can be reset after danger passes!

17. Some buildings in earthquake zones are designed to twist slightly during shaking. This twisting helps distribute the earthquake's force more evenly. It's like a building doing a gentle twist dance to stay safe during a quake!

18. Imagine a building that can change its weight distribution during an earthquake! Some smart buildings can quickly move heavy weights inside them to counteract swaying. It's like a building that can shift its body weight to keep its balance!

19. In New Zealand, engineers have developed buildings that rock during earthquakes. They lift slightly off their foundations and then settle back down. It's like giving a building permission to do a little hop when the ground shakes!

20. Some modern buildings use computer simulations to test thousands of earthquake scenarios before they're built. This helps engineers design the safest possible structures. It's like letting a building practice for earthquakes in a video game before facing the real thing!

Chapter 8: Solar-powered airplanes

1. Imagine an airplane that can fly day and night without any fuel! The Solar Impulse 2 did just that. It flew around the world using only the power of the sun. Its wings were covered in solar panels, like a flying carpet made of sunlight catchers!

2. Did you know there's a solar plane that can fly higher than regular airplanes? It's called the Helios Prototype. It can reach heights where the air is so thin, it's almost like space. It's like a solar-powered spider climbing an invisible web to the edge of our atmosphere!

3. Some solar planes are so light, they weigh less than a car! The Sunseeker II weighs only 280 pounds - that's about as much as a baby elephant. Its wings are long and thin, like a giant dragonfly powered by the sun.

4. Solar planes can be super quiet. The e-Genius, a solar-electric plane, makes less noise than a whisper when it flies overhead. Imagine an airplane so quiet, you could have a conversation while it passes by!

5. The Zephyr solar plane can stay in the air for weeks at a time! It's like a high-flying robot that never needs to land. During the day, it uses solar power to fly and charge its batteries. At night, it uses the stored energy to keep flying.

6. Some solar planes look like flying wings. The Solar Impulse has a wingspan as wide as a jumbo jet, but its body is tiny. It's like a giant boomerang covered in solar panels, silently gliding through the sky.

7. Engineers are working on solar planes that could explore other planets! These planes could fly in the thin atmosphere of Mars, powered by the distant sun. Imagine being the first to pilot a solar plane on another world!

8. The Sky Sailor is a tiny solar plane that weighs less than a gallon of milk. It's so light and efficient that it can fly using the power of a single light bulb! It's like a paper airplane that never needs to land.

9. Some solar planes use special lightweight batteries to store energy for nighttime flying. These batteries are so advanced, they're like tiny power plants that weigh almost nothing. It's like carrying a sun in your pocket!

10. The Solar Stratos plane is designed to fly so high, the pilot needs to wear a spacesuit! It can reach the edge of space using only solar power. It's like riding a ray of sunlight to the stars!

11. There's a solar plane called the Elektra One Solar that can be controlled from the ground like a remote-control toy. But this toy can fly for hours using just the power of the sun! Imagine playing with a life-sized solar-powered toy plane!

12. The Sunseeker Duo is a two-seater solar plane. It's like a flying tandem bicycle powered by the sun! The pilots sit one behind the other in a slim cockpit, soaring silently through the clouds.

13. Some solar planes use flexible solar panels that can bend with the shape of the wings. These panels are so thin and light, they're like a second skin for the airplane. It's as if the plane is wearing a solar-powered superhero suit!

14. Engineers are working on solar planes that could stay in the air for years! These planes would fly high above the clouds, acting like satellites. Imagine a plane that becomes your eye in the sky, never needing to come down!

15. The Silent 75 is a solar-powered electric plane that can take off and land on water. It's like a seaplane that drinks sunlight instead of fuel. You could hop between islands without ever needing to find a gas station!

16. Some solar planes use special propellers that can change their shape in flight. This helps them fly efficiently at different speeds. It's like having a Swiss Army knife for a propeller, always ready with the right tool for the job!

17. The SolarStratos plane has a pressurized cabin, like a tiny spaceship. The pilot breathes oxygen from a tank, just like a scuba diver. It's like scuba diving in the sky, powered by the sun!

18. Engineers are designing solar planes with wings that can change shape during flight. They spread wide to catch more sun during the day and fold in at night to fly more efficiently. It's like a bird that can shape-shift to follow the sun!

19. Some solar planes use special paint that helps them absorb more sunlight. This paint is so dark, it's like the plane is wearing a black hole! It turns every bit of sunlight into power for flying.

20. The Raymundo solar plane is designed to teach kids about solar power. It's small enough to fly in a gym and easy to build. Imagine having a science class where you get to fly your own solar plane!

Chapter 9: Hyperloop transportation

1. Imagine traveling in a pod that zooms through a tube at airplane speeds! That's what the Hyperloop is all about. It's like being in a super-fast train, but instead of wheels, it floats on air! You could travel from Los Angeles to San Francisco in just 30 minutes. That's faster than flying!

2. The Hyperloop tubes are designed to be almost completely airless inside. It's like creating a vacuum cleaner bag big enough for people to travel in! With almost no air resistance, the pods can go super fast without using much energy. It's like sledding down a hill that never ends!

3. Some Hyperloop designs use powerful magnets to make the pods float. It's called magnetic levitation, or maglev for short. Imagine riding in a vehicle that hovers above the track, as if by magic! It's like being in a real-life magic carpet ride.

4. Hyperloop pods might have virtual windows instead of real ones. These screens would show you what's outside as you zoom by. You could even choose to see different views! It's like being in a video game while you travel.

5. Engineers are working on ways to make Hyperloop pods change shape slightly as they move. This would help them slip through the air more easily, like a fish swimming through water. Imagine riding in a vehicle that wiggles like a snake to go faster!

6. Some people think Hyperloop stations could be built underground or even underwater! You might enter a station in the middle of a city and emerge on the other side of an ocean just hours later. It's like taking a secret tunnel to the other side of the world!

7. Hyperloop pods might be able to switch between different tubes, like cars changing lanes on a highway. This would let you travel to many different cities without ever leaving your pod. It's like being in an elevator that can go sideways as well as up and down!

8. The Hyperloop could be powered by solar panels on top of the tubes. It would be like riding in a giant solar-powered straw! This clean energy could make the Hyperloop one of the most eco-friendly ways to travel long distances.

9. Some designers want to make Hyperloop pods that can drive on regular roads too. You could zoom through the tube for long distances, then drive right to your destination. It's like having a car that turns into a super-fast spaceship!

10. Hyperloop tubes might be built on tall pylons above the ground. This would let them go over hills and valleys in a straight line. Riding in one would be like flying, but inside a tube! You could see amazing views as you speed along high above the ground.

11. Engineers are working on ways to make Hyperloop safe from earthquakes. The tubes might be designed to move and flex during a quake. It would be like riding in a giant, flexible straw that can bend without breaking!

12. Some Hyperloop designs include entertainment systems in the pods. You could watch movies, play games, or even have a virtual reality experience while you travel. It's like having your own private amusement park ride that takes you to your destination!

13. Hyperloop pods might use special air bearings to float. These work by blowing out tiny jets of air, creating a cushion to ride on. It's like hovering on a bunch of tiny hover boards all working together!

14. Scientists are exploring ways to use Hyperloop for cargo transportation too. Imagine your online shopping order zooming through a tube at nearly the speed of sound! It would be like having a super-fast pneumatic tube, like at a bank drive-through, but big enough for packages.

15. Some people think Hyperloop stations could be like small airports, with shops and restaurants. You could have a meal or do some shopping while waiting for your pod. It's like a shopping mall that sends you on an adventure when you're done!

16. Hyperloop pods might have special brakes that use magnets instead of friction. These would help the pod slow down quickly and safely. It's like having invisible hands gently squeezing the pod to make it stop.

17. Engineers are working on ways to pressurize Hyperloop pods, like airplanes. This would keep passengers comfortable even when traveling through low-pressure tubes. It's like being in a cozy bubble as you zoom through space!

18. Some Hyperloop designs include emergency exits along the tube. If a pod needed to stop, passengers could exit safely. It's like having secret doors in a giant, high-tech tunnel!

19. Hyperloop might use artificial intelligence to control the pods. This smart system could adjust speed and route for the smoothest, fastest trip. It would be like having the world's smartest driver taking you on a super-speed adventure!

20. Scientists are even thinking about using Hyperloop technology for space travel! Imagine zooming through a tube to launch into orbit. It could be like a giant slingshot to send spacecraft into space without rockets!

Chapter 10: Vertical farming technology

1. Imagine a farm that goes up instead of out! That's what vertical farming is all about. In Singapore, there's a vertical farm that looks like a giant bookshelf full of veggies. Each shelf has its own water and light, so plants can grow all year round. It's like having a skyscraper made of lettuce!

2. Did you know some vertical farms use pink lights? Plants love this special light - it helps them grow super fast. Walking into one of these farms is like stepping into a magical pink world where veggies grow in glowing towers. It's as if the plants are having their own disco party!

3. Some vertical farms use no soil at all! Instead, plant roots dangle in nutrient-rich water. This method is called hydroponics. It's like the plants are swimming and eating at the same time! These farms use way less water than regular farms and can grow food almost anywhere.

4. Vertical farms can grow food in the middle of busy cities. Imagine having fresh tomatoes grown right next door to your apartment! Some restaurants even have their own vertical farms, so your salad is picked just minutes before you eat it. It's like having a garden in your fridge!

5. In Japan, there's a vertical farm in an old factory. Robots do a lot of the work, planting seeds and harvesting crops. It's like having a team of robot farmers working 24/7 to grow your food. These robots never get tired and can work in the dark!

6. Some vertical farms use fish to help grow plants! The fish live in tanks below the plants, and their waste fertilizes the crops. In return, the plants clean the water for the fish. It's called aquaponics, and it's like having a farm and an aquarium in one!

7. Vertical farms can grow food in some pretty weird places. There's even a plan to build one in Antarctica! Imagine eating fresh lettuce at the South Pole, grown in a cozy indoor farm while it's freezing outside. It's like having a little piece of summer in the middle of a snowy wasteland!

8. In Wyoming, there's a vertical farm inside shipping containers. These can be stacked up like giant Lego bricks to make a farm of any size. Each container can grow as much food as an acre of regular farmland! It's like having a magic box that's bigger on the inside.

9. Some vertical farms use special LEDs that change color as plants grow. The lights start blue when plants are young, then turn red as they get older. It's like the plants have mood lighting that changes as they grow up!

10. Vertical farms can grow more than just veggies. Some are growing flowers, and others are even trying to grow tiny trees! Imagine a tower full of apple trees, each one growing sideways out of the wall. It's like a fairytale forest turned on its side!

11. In New Jersey, there's a vertical farm that grows over 2 million pounds of greens each year! That's as heavy as 200 elephants! All of this food comes from a building about the size of a soccer field. It's like having an entire county's worth of farms squeezed into one big building.

12. Some vertical farms use special robots that look like shelves on wheels. These robots move plants around to give them the perfect amount of light and water. It's like having a hotel for plants where robot butlers take care of their every need!

13. Vertical farms can grow food really fast. Some leafy greens are ready to eat just 10 days after planting! That's way quicker than in a regular farm. It's like having a fast-forward button for growing vegetables.

14. In Scotland, scientists are using vertical farming to grow medicine. They're growing special plants that can be turned into treatments for diseases. It's like having a hospital and a farm combined into one high-tech tower!

15. Some people are designing vertical farms that float on the ocean! These would be like giant greenhouses bobbing on the waves, growing food for coastal cities. Imagine sailing past a farm on your way to the beach!

16. Vertical farms can help save endangered plants. Scientists can grow rare species in perfect conditions, protected from threats in the wild. It's like building a super-safe hotel for plants that are in danger of disappearing.

17. In Dubai, they're planning to build a vertical farm that looks like a giant beehive! The hexagon-shaped levels will be stacked up high, filled with plants. It's like creating a skyscraper-sized home for bees, but filled with vegetables instead of honey.

18. Some vertical farms use special sensors to listen to plants! These sensors can tell when plants need water or nutrients by detecting tiny vibrations. It's like the plants are whispering what they need, and the farmers have super-hearing to understand them.

19. Vertical farms can grow food in space! NASA is experimenting with growing vegetables on the International Space Station. Imagine being an astronaut and eating a fresh salad that was grown right next to you in zero gravity!

20. In Malaysia, there's a vertical farm that uses old plastic bottles as planters. They hang from ceiling to floor, creating walls of greenery. It's like recycling and farming got together to make a beautiful, edible curtain!

Chapter 11: 3D-printed organs

1. Imagine printing a new ear for someone who needs one! Scientists can now use special 3D printers to make ear shapes using a gel made from living cells. Over time, these cells grow into real ear tissue. It's like printing a seed that grows into an ear!

2. Did you know doctors can print tiny hearts the size of a cherry? These mini-hearts beat just like real ones! They're too small for transplants, but they help scientists study heart diseases and test new medicines. It's like having a tiny drum set that plays the rhythm of life!

3. Some scientists are working on printing skin for burn victims. They use a 3D printer that's a bit like a color printer, but instead of ink, it uses different types of skin cells. Imagine having a printer that could make a band-aid that turns into real skin!

4. Researchers have printed a human cornea - the clear front part of the eye. They used a special bio-ink made from stem cells. It's like printing a tiny contact lens made of living cells that could help someone see again!

5. Scientists in Israel printed a tiny heart with blood vessels. It was the size of a rabbit's heart and made from human cells. Imagine holding a heart in your hand that was made by a printer! It's like science fiction becoming real.

6. Doctors have used 3D printing to make custom bones for patients. They scan the person's body and print an exact replacement bone. It's like having a spare parts shop for your skeleton, where each piece is made just for you!

7. Some researchers are trying to print livers. The liver is super complex, with lots of different cell types. Printing one is like trying to print a whole factory! Scientists hope these printed livers could one day help people who need liver transplants.

8. Did you know scientists can print blood vessels? They're like tiny pipes that carry blood in our bodies. Printing them is tricky because they're so small. It's like trying to print spaghetti that can carry tomato sauce!

9. Researchers are working on printing pancreases to help people with diabetes. The printed pancreas would make insulin, just like a real one. Imagine having a tiny insulin factory printed just for you!

10. Some scientists are trying to print kidneys. Kidneys are like the body's cleaning system, filtering out waste. Printing a kidney is like trying to print a super-complicated water filter. It's tricky, but scientists are working hard to figure it out!

11. Doctors have used 3D printing to make models of patients' hearts before surgery. This helps them plan exactly what to do. It's like having a practice heart to work on before the big game!

12. Scientists are experimenting with printing cartilage - the bendy stuff in your nose and ears. They mix cartilage cells with a special gel and print ear and nose shapes. It's like printing with living play-doh!

13. Some researchers are trying to print lungs. Lungs are super complex, with millions of tiny air sacs. Printing them is like trying to print a tiny forest of air bubbles! It's very difficult, but scientists are working hard on it.

14. Did you know scientists can print tiny bits of brain tissue? They're too small to think, but they help researchers study brain diseases. It's like printing a tiny piece of a super-computer to see how it works!

15. Doctors have used 3D printing to make custom skull implants for patients. They scan the person's head and print a piece that fits perfectly. It's like printing a custom helmet that goes inside your head!

16. Some scientists are working on printing stomachs. They hope these could help people with stomach diseases. Imagine printing an organ that could digest your food for you!

17. Researchers have printed tiny bits of spinal cord tissue. This could one day help people with spinal cord injuries. It's like printing a tiny piece of an electrical cable that sends messages from your brain to your body!

18. Scientists are experimenting with printing muscle tissue. They mix muscle cells with a special gel and print them in layers. It's like printing a piece of steak, layer by layer!

19. Some researchers are trying to print thyroid glands. These glands help control how your body uses energy. Printing one is like trying to print a tiny factory that makes important hormones!

20. Did you know scientists can print bone marrow? Bone marrow is where your blood cells are made. Printing it is like trying to print a tiny blood cell factory! This could one day help people with blood diseases.

Chapter 12: Robotic exoskeletons

1. Imagine wearing a robot suit that gives you super strength! That's what a robotic exoskeleton does. Some can help people lift really heavy things without getting tired. It's like having an Iron Man suit in real life!

2. There's a special exoskeleton that helps people walk again after injuries. It has motors at the hips and knees that move your legs for you. It's like having a friendly robot giving you a piggyback ride while you walk!

3. Scientists have made an exoskeleton that can read your mind! It uses sensors on your head to know when you want to move. Then it moves your arm for you. It's like having a robot friend who always knows what you're thinking!

4. Some factories use exoskeletons to help workers lift heavy things. These suits support the workers' backs and arms, so they don't get tired or hurt. It's like giving every worker a pair of super-strong robot arms!

5. There's an exoskeleton that helps kids with muscle problems walk. It's like wearing a pair of robot legs that help you take each step. Imagine having a robot buddy that helps you play and run around!

6. Scientists have made an exoskeleton hand that can crush cans like they're made of paper! It makes your hand super strong. It's like having a robot glove that gives you the strength of a giant!

7. Some exoskeletons are so light, you can wear them under your clothes. They're made of soft materials that move with your body. It's like wearing a second skin that gives you secret super powers!

8. There's an exoskeleton that helps firefighters carry heavy equipment. It makes their gear feel light as a feather. Imagine being a superhero firefighter with robot muscles to help save the day!

9. Scientists have made an exoskeleton that helps you run faster. It has springs that store energy when you step and release it to push you forward. It's like having robot springs in your shoes that make you zoom along!

10. Some exoskeletons can help you stand for a long time without getting tired. They have a seat that folds out when you need it. It's like carrying your own comfy chair wherever you go!

11. There's an exoskeleton that helps surgeons perform operations. It steadies their hands and helps them make super precise movements. It's like having a robot assistant that makes you the world's steadiest surgeon!

12. Scientists have made an exoskeleton that helps you lift things with just one finger! It multiplies the strength of your finger by 40 times. Imagine being able to lift a heavy box with just your pinky!

13. Some exoskeletons are designed for outer space. They help astronauts exercise in zero gravity to keep their muscles strong. It's like having a portable gym that works even when you're floating!

14. There's an exoskeleton that helps people with weak arms eat by themselves. It supports their arm and helps them guide the spoon to their mouth. It's like having a friendly robot arm that helps you enjoy your favorite meals!

15. Scientists have made an exoskeleton that helps you swim faster. It has fins that move with your legs to push you through the water. Imagine swimming like a mermaid with robot fins!

16. Some exoskeletons are made to help elderly people stay active. They support the body and make it easier to walk and climb stairs. It's like giving grandma and grandpa super powers to keep up with the grandkids!

17. There's an exoskeleton that helps soldiers carry heavy backpacks. It transfers the weight to the ground so their backs don't get tired. Imagine wearing a backpack that feels as light as a feather, no matter how full it is!

18. Scientists have made an exoskeleton that helps you jump really high. It has powerful springs in the legs that launch you into the air. It's like having pogo sticks built into your shoes!

19. Some exoskeletons are designed to help people with hand tremors. They steady the hand so people can write and eat more easily. It's like having a gentle robot hand guiding your own to keep it still!

20. There's an exoskeleton that helps workers paint ceilings without getting tired arms. It supports their arms as they work above their heads. Imagine being able to paint like Michelangelo without your arms ever getting tired!

Chapter 13: Space elevators

1. Imagine riding an elevator all the way to space! That's what a space elevator would do. It's like a super tall building that reaches from Earth to way above the clouds. Instead of rockets, we could just ride up to space in a special car!

2. Space elevators would use a special rope stronger than steel but lighter than plastic. Scientists call it a carbon nanotube ribbon. It's so strong, it could hold up the entire elevator cable, which would be thousands of miles long. That's like using a piece of string to hold up a skyscraper!

3. The top of a space elevator would be in orbit around Earth. It would spin with our planet, always staying above the same spot. It's like having a really tall flagpole that reaches all the way to space, with the top moving in a big circle every day.

4. Riding a space elevator would take about a week to reach space. That's much slower than a rocket, but way more comfortable! Imagine spending a week in a climbing elevator, watching Earth get smaller and smaller below you.

5. Space elevators could make space travel much cheaper. Instead of using expensive rockets, we could just ride the elevator up! It would be like taking a bus to space instead of a super-fast, super-expensive race car.

6. The base of a space elevator would probably be on a platform in the ocean near the equator. This is the best place to build it because the Earth spins fastest here. It's like picking the perfect spot to build the world's tallest treehouse!

7. Space elevators could help us build hotels in space! We could send up building materials easily and construct fancy space hotels. Imagine having a summer vacation in orbit, looking down at Earth from your hotel room window!

8. Scientists think space elevators could help us clean up space junk. We could use them to bring down old satellites and rocket parts that are floating around Earth. It's like using a really tall grabber tool to pick up litter in space!

9. Some people think we should build a space elevator on the Moon first. It would be easier because the Moon's gravity is weaker. Imagine riding an elevator from the Moon's surface all the way to a space station orbiting the Moon!

10. Space elevators could help us send supplies to astronauts more easily. Instead of launching rockets full of food and water, we could just send it up the elevator. It would be like having the world's tallest dumbwaiter to deliver snacks to space!

11. The cars that ride up the space elevator would probably use electricity to climb. They might have solar panels to catch energy from the sun as they go up. Imagine riding in a solar-powered space taxi that climbs a giant rope to the stars!

12. Building a space elevator would be a huge project. It would take thousands of people working together and might cost billions of dollars. But once it's built, it could change how we explore space forever! It's like building a bridge to the stars.

13. Space elevators could help us study Earth better. We could easily send up lots of cameras and sensors to watch our planet from above. It would be like giving Earth a bunch of high-tech friendship bracelets that help us understand it better!

14. Some scientists think we could use space elevators to launch spacecraft to other planets. The spacecraft could start from the top of the elevator, already in space! It's like giving spaceships a big head start on their journey.

15. The idea of space elevators isn't new - a Russian scientist thought of it over 100 years ago! But we're only now getting close to having the technology to build one. It's like a 100-year-old dream that might finally come true!

16. Space elevators could help us build giant solar panels in space. These could collect lots of energy from the sun and send it back to Earth. Imagine having a power plant in space that gives clean energy to the whole world!

17. Some people worry that space debris might hit the elevator cable. To protect it, we might need to use lasers to zap away dangerous space junk. It's like having a high-tech shield to protect our cosmic climbing rope!

18. The top of a space elevator would be so high up, you could see the curve of the Earth and the blackness of space. Imagine riding to the top and seeing our whole planet spread out below you like a giant blue marble!

19. Space elevators might have emergency escape pods, just in case something goes wrong. These pods could safely bring people back to Earth. It's like having a cosmic fire escape for your space adventure!

20. Some scientists think we could use space elevators to bring valuable materials back from asteroids. We could mine asteroids for rare metals and bring them down the elevator. It's like having a cosmic mining operation with the world's longest conveyor belt!

Chapter 14: Invisibility cloaks and metamaterials

1. Imagine wearing a jacket that makes you invisible! Scientists are working on special materials called metamaterials that can bend light around objects. It's like wearing a magic cloak that makes light go around you instead of bouncing off you, so no one can see you!

2. Did you know some animals have natural invisibility cloaks? Octopuses can change their skin color and texture to blend in with their surroundings. Scientists are trying to make materials that can do the same thing. It's like creating a high-tech octopus skin!

3. Metamaterials aren't just for invisibility. Some can make things super quiet! They absorb sound waves, making noisy things silent. Imagine having a bedroom wall that could make the noise from a busy street disappear like magic!

4. Scientists have made a material that can hide things from earthquakes! It redirects the earthquake waves around buildings, keeping them safe. It's like having an invisible force field that protects houses from nature's shakes and rumbles!

5. There's a special paint that can make things invisible to infrared cameras. These cameras see heat, but this paint hides heat signatures. It's like having camouflage that works against robot eyes!

6. Some metamaterials can make things invisible to radar. This could help make airplanes that don't show up on radar screens. It's like giving planes their own stealth superpowers!

7. Imagine a tablecloth that could make everything under it disappear! Scientists have created small versions of this using metamaterials. It's not perfect yet, but it's like having a real-life magic trick at your dinner table!

8. Metamaterials can also be used to make super-thin lenses. These lenses could help make smaller, more powerful cameras and telescopes. It's like shrinking a big magnifying glass down to the size of a contact lens!

9. Some researchers are working on metamaterials that can hide things underwater. This could help protect ships and submarines from sonar. It's like giving boats an invisibility cloak that works in the ocean!

10. Did you know scientists have made a material that can hide time? It can make a short pause in a light beam, creating a brief moment of invisibility. It's like having a remote control that can pause real life for a split second!

11. Metamaterials might one day help us see through walls! They could be used to make devices that can "cancel out" the effect of walls on light or radio waves. It's like giving people X-ray vision without the harmful radiation!

12. Some metamaterials can make things appear smaller than they really are. This could be used to hide big objects or make more space inside things like cars or airplanes. It's like having a shrink ray that only affects how things look!

13. Imagine a roof that can make your whole house invisible! While we can't do this yet, scientists are working on metamaterials that could one day make large objects disappear. It's like turning your home into a chameleon!

14. Metamaterials can also be used to make things that are usually invisible become visible. This could help doctors see tiny things inside our bodies more clearly. It's like having superhero vision that can spot hidden problems!

15. Some researchers are trying to make "acoustic cloaks" that can hide objects from sonar and other sound-based detection methods. It's like having an invisibility cloak for your ears instead of your eyes!

16. Did you know there are metamaterials that can make heat flow backwards? Usually, heat moves from hot things to cold things, but these materials can reverse that. It's like having a magic wand that can make ice cubes boil water!

17. Scientists have created a metamaterial that can make light move in a spiral. This could be used to make better optical computers in the future. It's like giving light a roller coaster track to follow inside a computer chip!

18. Some metamaterials can make things appear to levitate. They bend light in a way that makes objects look like they're floating above where they really are. It's like having a magic trick built into the material itself!

19. Researchers are working on metamaterials that can change their properties when you stretch them. This could lead to clothes that change color or pattern when you move. Imagine having a t-shirt that puts on a light show every time you dance!

20. There are even metamaterials that can control earthquakes! They can redirect seismic waves around important buildings or areas. It's like having an invisible traffic cop for earthquake waves, telling them where they can and can't go!

Chapter 15: Maglev trains

1. Imagine a train that floats in the air! That's what maglev trains do. They use powerful magnets to lift off the tracks and zoom along without touching them. It's like riding on a magic carpet made of science!

2. The world's fastest maglev train is in China. It can go over 370 miles per hour! That's faster than a race car or even some airplanes. You could travel from New York to Washington D.C. in less than an hour on a train this fast!

3. Maglev trains are super quiet. Because they don't touch the tracks, there's no clickety-clack sound like regular trains. It's more like gliding through the air. Imagine riding in a train that's as quiet as a whisper!

4. In Japan, engineers are working on a maglev train that will travel through tunnels in the mountains. It will be like riding in a super-fast underground rocket ship! The train will zip between cities faster than you can say "Tokyo to Nagoya!"

5. Maglev trains don't need engines like regular trains. They're pushed along by magnets on the track. It's like having an invisible giant pushing your train car really fast!

6. Some scientists think we could one day build maglev trains that go through vacuum tubes. With no air resistance, these trains could go incredibly fast - maybe even faster than airplanes! It would be like traveling in a super-speed pipeline.

7. Maglev trains are great for the environment. They don't use gasoline and don't make pollution like cars or planes. Riding a maglev train is like giving the Earth a big, green hug!

8. The first commercial maglev train was built in Shanghai, China. It takes people from the airport to the city in just 7 minutes! That's like teleporting from the airplane to your hotel.

9. Maglev trains can go up steep hills more easily than regular trains. The magnets give them super climbing powers! It's like having a train that thinks it's a mountain goat.

10. Some people have suggested building maglev trains that go around the whole world! Imagine being able to take a train from New York to London, across the ocean. It would be like a global roller coaster ride!

11. Maglev trains don't wear out the tracks like regular trains do. The tracks last longer because the train never touches them. It's like having shoes that never get holes because they float above the ground!

12. In Germany, there's a maglev train that's more like a monorail. It hangs from the track instead of floating above it. It's like riding in a high-tech banana that's being carried by a giant magnetic hand!

13. Some scientists are working on personal maglev pods. These would be like tiny maglev trains for just one or two people. Imagine having your own personal floating car to take you to school!

14. Maglev trains can start and stop really quickly. They don't need time to speed up or slow down like regular trains. It's like riding a train that has super-fast reflexes!

15. In South Korea, there's a maglev train in an amusement park. It's not very fast, but it shows people how cool maglev technology is. It's like a science lesson and a fun ride all in one!

16. Some people think we could use maglev technology to launch spacecraft. The spacecraft would ride a maglev track to get up to speed before taking off. It would be like giving a rocket a super-fast running start!

17. Maglev trains are very stable. Even at high speeds, they don't shake or rattle. Riding in one feels as smooth as gliding on ice. It's like traveling on a perfectly straight, invisible road.

18. Engineers are working on ways to make maglev trains even when the power goes out. They're designing special emergency wheels that can pop out if needed. It's like giving the train a backup pair of roller skates!

19. Some maglev train designs look like bullets or airplane wings. This shape helps them slip through the air easily. It's like turning a train into a giant, flying arrow!

20. In the future, we might have maglev networks in cities, with small pods that can take you exactly where you want to go. It would be like having a whole city connected by invisible, floating roads!

Chapter 16: Artificial intelligence in robotics

1. Meet Pepper, a friendly robot that can read human emotions! Pepper uses AI to understand how people are feeling by looking at their faces and listening to their voices. It's like having a robot friend who always knows when you need a cheer-up or a high-five!

2. In Japan, there's a hotel where robot dinosaurs check you in! These dinos use AI to understand what guests are saying and help them get to their rooms. Imagine being greeted by a T-Rex that can speak your language!

3. Scientists have made a robot called Sophia that can have conversations with people. Sophia uses AI to understand questions and come up with answers. It's like talking to a computer that has its own thoughts and ideas!

4. There's a robot called Atlas that can do backflips! It uses AI to balance and move its body, just like a gymnast. Imagine having a robot buddy that could teach you awesome acrobatic tricks!

5. In some hospitals, tiny AI-powered robots help deliver medicine to patients. They can find their way around the hospital all by themselves! It's like having a smart little delivery truck bringing you what you need to feel better.

6. Scientists have created robots that can learn to walk by themselves, just like babies do! These robots use AI to figure out how to balance and move their legs. It's like watching a robot grow up super fast!

7. There's a robot called Moxi that helps nurses in hospitals. Moxi can deliver supplies and even give patients a friendly wave! It's like having a helpful robot assistant that never gets tired of lending a hand.

8. In some factories, robots use AI to work together like a team. They can figure out how to divide tasks and help each other out. It's like watching a robot sports team, but instead of playing games, they're building cool stuff!

9. Scientists have made a robot fish that can swim in the ocean! It uses AI to move its tail and fins just like a real fish. Imagine having a robot pet that could explore the sea with you!

10. There's a robot called Spot that looks like a dog and can climb stairs and open doors! Spot uses AI to see obstacles and figure out how to move around them. It's like having a robot puppy that's super smart and helpful!

11. In some restaurants, robot chefs use AI to cook food! They can learn new recipes and even create their own. Imagine having a robot that could make you the perfect pizza every time!

12. Scientists have created tiny robots that can work together to build things. These "swarm robots" use AI to communicate and coordinate their actions. It's like watching a team of ant-sized construction workers building a tiny city!

13. There's a robot called Asimo that can run, climb stairs, and even dance! Asimo uses AI to move smoothly and keep its balance. Imagine having a robot friend who could be your dance partner at every party!

14. In some warehouses, robots use AI to find and pack items for shipping. They can zoom around on wheels and pick up boxes with their robot arms. It's like having a team of super-fast, super-strong helpers getting your online orders ready!

15. Scientists have made robots that can learn to play games just by watching humans. These robots use AI to understand the rules and come up with strategies. Imagine having a robot buddy that could learn any game you teach it!

16. There's a robot called Flippy that can flip burgers in restaurants! Flippy uses AI to know when the burgers are cooked just right. It's like having a robot chef that makes perfect burgers every time!

17. In some homes, robot vacuums use AI to clean floors all by themselves. They can remember the layout of your house and avoid obstacles. Imagine having a little robot helper that keeps your room clean without you lifting a finger!

18. Scientists have created robots that can express emotions using AI. These robots can smile, frown, and even look surprised! It's like having a robot friend that can play charades with its face.

19. There's a robot called Kuri that can be a home companion. Kuri uses AI to recognize family members, play music, and even take photos. It's like having a robot pet that's also your personal DJ and photographer!

20. In some schools, robot tutors use AI to help kids learn. These robots can adapt their teaching style to each student's needs. Imagine having a robot teacher that always knows exactly how to explain things so you understand!

Chapter 17: Bionics and prosthetic limbs

1. Meet Sarah, who has a special bionic arm! It moves just like a real arm, but it's made of cool metals and plastics. Sarah can control it with her thoughts, just like a regular arm. She can even play video games with it! It's like having a superhero arm that listens to your brain.

2. Did you know there are prosthetic legs that can help people run super fast? Some athletes use these special legs in the Paralympics. They're shaped like curved blades and can make runners zoom along like speedy cheetahs!

3. Scientists have made a bionic hand that can feel things! It has special sensors that send signals to the brain, so people can feel if something is soft or hard. Imagine having a robot hand that could tell you how fluffy your teddy bear is!

4. There's a boy named Cameron who has a prosthetic leg with cool LED lights. He can change the colors whenever he wants! It's like having a leg that's also a disco light show. Cameron says it makes him feel like a superhero.

5. Researchers have created prosthetic arms that can be controlled by muscle signals. When you think about moving your arm, tiny electrical signals in your muscles tell the prosthetic what to do. It's like having a mind-reading robot arm!

6. Some prosthetic hands can do really precise movements, like playing the piano! They have motors in each finger that can move independently. Imagine having a robot hand that could play beautiful music or do tricky puzzles.

7. There's a girl named Tilly who has a waterproof prosthetic leg. She can go swimming and play in the ocean without worrying about it getting damaged. It's like having a super leg that's part fish!

8. Scientists have made prosthetic eyes that can help blind people see! These bionic eyes send signals directly to the brain. It's not perfect vision, but it can help people see shapes and movement. Imagine having robot eyes that give you super sight!

9. Some prosthetic legs have built-in computers that help people walk more naturally. These smart legs can adjust to different terrains, like going up stairs or walking on sand. It's like having a leg that's also a tiny robot brain!

10. There's a man named Jason who has a prosthetic arm with a built-in flashlight, bottle opener, and even a USB port to charge his phone! It's like having a Swiss Army knife for an arm.

11. Researchers are working on prosthetic skin that can feel temperature and pressure. This could help people with prosthetic limbs feel things more like they would with real skin. Imagine having robot skin that could tell you if your ice cream was too cold!

12. Some prosthetic hands can be customized with cool designs. People have made hands that look like robot parts, or covered in colorful patterns. It's like having an arm that's also a piece of art!

13. There's a boy named Zion who was born without hands, but now he has two bionic hands! He can play basketball, write, and even make his own lunch. It's like having two robot helpers attached to your arms.

14. Scientists have made prosthetic legs that can help people climb rocks! These legs have special ankles that can move in all directions, making it easier to find footholds. Imagine having robot legs that could turn you into a super climber!

15. Some prosthetic arms have built-in tools, like screwdrivers or wrenches. People can switch out different attachments depending on what they need. It's like having a whole toolbox as part of your body!

16. There's a girl named Easton who has a prosthetic arm that she can decorate with stickers and drawings. She changes its look all the time to match her mood or outfit. It's like having an arm that's also a creative canvas!

17. Researchers are working on prosthetic limbs that can grow as children grow. These expandable limbs can be adjusted so kids don't need new ones as often. Imagine having a robot arm that grows with you!

18. Some prosthetic legs have shock absorbers built in, kind of like the ones in cars. This makes walking more comfortable and protects the rest of the body from jolts. It's like having a super-comfy, bouncy leg!

19. There's a man named Dan who has a prosthetic hand that can be controlled by an app on his phone. He can select different grip patterns for different tasks. It's like having a hand with its own remote control!

20. Scientists are developing prosthetic limbs that can send signals back to the brain, giving people a sense of touch and movement. Imagine having a robot arm that could actually feel things and tell your brain about it!

Chapter 18: Underwater habitats

1. Imagine living in a house under the sea! Scientists have built underwater habitats where people can stay for weeks. These bubble-like homes have bedrooms, kitchens, and even laboratories. It's like camping in a giant fishbowl!

2. Did you know there's an underwater hotel in Florida? Guests have to scuba dive to reach their rooms! You can sleep surrounded by fish and coral reefs. It's like having a sleepover in an aquarium, but you're the one in the tank!

3. In the 1960s, some scientists lived in an underwater lab called Sealab. They stayed underwater for weeks, studying ocean life. It was like having a secret clubhouse at the bottom of the sea!

4. Some underwater habitats are used to train astronauts! Living underwater is a bit like living in space - you're in a small, isolated place where it's hard to go outside. It's like practicing for a trip to Mars by visiting Atlantis!

5. There's an underwater lab called Aquarius off the coast of Florida. Scientists live there to study coral reefs. They can swim out of their front door right into the ocean! It's like having a treehouse, but for fish instead of birds.

6. Imagine a whole city under the sea! Some architects have designed underwater cities with homes, schools, and even parks. These futuristic cities could help if sea levels rise. It's like moving to Bikini Bottom from SpongeBob SquarePants!

7. In Japan, there's a plan to build underwater farms. These would grow fruits and vegetables in big clear domes under the sea. Imagine picking strawberries while fish swim by your window!

8. Some underwater habitats are used to study how people behave when they're isolated. Scientists watch how people work together when they're stuck in a small space underwater. It's like a reality TV show, but for science!

9. There's an idea to build underwater gas stations for submarines. These would let subs stay underwater longer without coming up for supplies. It's like having a pit stop for underwater race cars!

10. Imagine an underwater zoo where you visit sea animals in their natural habitat! Some scientists have proposed building observation domes on the sea floor. You could watch whales and sharks swim by your window!

11. In the future, we might have underwater factories. These could use the ocean's resources to make things without polluting the land. It's like having a secret workshop hidden under the waves!

12. Some people think we should build underwater libraries to protect books from disasters on land. These water-proof vaults would keep knowledge safe. It's like having a treasure chest of stories at the bottom of the sea!

13. Imagine an underwater train that connects continents! Some engineers have proposed building clear tunnels across oceans. You could travel from America to Europe while watching fish swim by. It's like a subway ride through a giant aquarium!

14. There's an idea to build underwater greenhouses. These would use the sun's light filtering through the water to grow plants. Imagine having a garden where seaweed and tomatoes grow side by side!

15. Some scientists want to build underwater labs to study deep-sea creatures. These labs would be like space stations, but on the ocean floor. You could watch bizarre glowing fish right outside your window!

16. Imagine an underwater museum! Some artists have created sculptures that are placed on the ocean floor. Divers can swim through these underwater art galleries. It's like having a treasure hunt in a sunken city!

17. There's a plan to build underwater hotels that move around the ocean. These floating hotels would drift with the currents, giving guests a new view every day. It's like going on a cruise, but you're under the waves instead of on top!

18. Some engineers are designing underwater repair stations for ships. These would be like underwater garages where boats could get fixed without going to shore. Imagine being a mechanic in a giant bubble under the sea!

19. There's an idea to build underwater power plants that use the ocean's currents to make electricity. These plants would be like giant underwater windmills. It's like harnessing the power of rivers, but in the open ocean!

20. Imagine an underwater research station that can dive deep into the ocean! This mobile lab would be like a submarine and a house combined. Scientists could explore the deepest parts of the sea while living comfortably. It's like having a camper van that can visit the Mariana Trench!

Chapter 19: Smart dust and micro-sensors

1. Imagine tiny robots as small as dust particles! That's what smart dust is. Scientists can scatter these tiny sensors to collect information about the environment. It's like having a team of invisible helpers that can tell you all about the world around you!

2. Did you know there are sensors so small they can fit inside a raindrop? These micro-sensors can measure things like temperature and humidity. It's like giving each raindrop its own tiny weather station!

3. Some farmers use smart dust to watch their crops. The tiny sensors can tell if plants need water or if pests are attacking. It's like having a bunch of microscopic farmers keeping an eye on every single plant!

4. Scientists have made sensors that can float in the air like pollen. These could help track air pollution or monitor the weather. Imagine a cloud of tiny robot explorers floating on the wind, sending back information about the sky!

5. There are micro-sensors that can be put into paint! This "smart paint" can detect cracks in buildings or bridges. It's like giving walls and bridges the ability to tell us when they need fixing!

6. Some doctors are working on micro-sensors that can travel through your body. These tiny robots could help find diseases or deliver medicine. It's like having a team of microscopic doctors exploring inside you!

7. Imagine smart dust that can detect earthquakes! Scientists could spread these tiny sensors in areas prone to quakes. It would be like giving the Earth its own network of mini seismographs.

8. There are micro-sensors that can be woven into clothes. These "smart clothes" could monitor your health or even change color with your mood. It's like wearing a t-shirt that's also a doctor and a mood ring!

9. Some cities use smart dust to monitor traffic. Tiny sensors on the roads can count cars and detect traffic jams. It's like having an army of mini traffic cops watching every street!

10. Scientists have made sensors small enough to stick on a bee's back! These help researchers study how bees behave and why some bee populations are shrinking. It's like giving each bee its own tiny backpack full of scientific instruments.

11. Imagine smart dust that can detect forest fires before they get big! These sensors could smell smoke or feel heat, alerting firefighters early. It's like having a bunch of tiny fire watchers scattered throughout the forest.

12. There are micro-sensors that can be put into food packaging. These can tell you if food has gone bad before you even open it. It's like having a tiny food taster checking every bite for you!

13. Some scientists use smart dust to study volcanoes. The tiny sensors can detect gases and tremors that might mean an eruption is coming. It's like giving a volcano its own early warning system made of dust!

14. Imagine micro-sensors that can be sprayed on your skin like sunscreen! These could monitor your sun exposure and tell you when it's time to reapply or go into the shade. It's like having a tiny lifeguard watching your skin all day.

15. There are smart dust particles that can clean up oil spills. These tiny robots can absorb oil and then be collected, helping to clean the ocean. It's like having a swarm of microscopic janitors cleaning up the sea!

16. Scientists have made sensors small enough to fit inside a snowflake! These can help study how snow forms and moves. It's like giving each snowflake its own tiny diary to record its journey.

17. Imagine smart dust that can help find lost hikers in the woods. These sensors could detect movement or hear calls for help. It's like having a forest full of tiny search and rescue workers always on duty!

18. There are micro-sensors that can be put into car tires. These can tell drivers when the tire pressure is low or if there's a puncture. It's like giving each tire its own health check-up system!

19. Some researchers use smart dust to study animal migrations. Tiny sensors on birds or fish can track their movements across the globe. It's like giving each animal its own miniature GPS and travel diary!

20. Imagine smart dust that can help grow plants on Mars! These tiny sensors could monitor soil conditions and help astronauts farm on other planets. It's like having a team of microscopic gardeners preparing for life on Mars!

Chapter 20: Fusion reactors

1. Imagine a star in a bottle! That's what a fusion reactor is like. Scientists are trying to recreate the power of the sun right here on Earth. They heat up special gases until they're hotter than the sun's surface. It's like having a mini-star that could power whole cities!

2. Did you know that fusion is what makes stars shine? In a fusion reactor, tiny particles slam together and release huge amounts of energy. It's like billions of atomic firecrackers going off at once, but instead of noise, they make power!

3. Fusion reactors use super strong magnets to control the hot gas. These magnets are so powerful they could lift a car! They keep the hot gas floating in the middle of the reactor, never touching the walls. It's like playing the world's most extreme game of "don't touch the floor"!

4. Scientists use lasers in some fusion experiments. These aren't your ordinary laser pointers - they're some of the most powerful lasers in the world! They heat the fuel up to millions of degrees. It's like using a lightsaber to start a campfire!

5. Fusion reactors could one day use seawater as fuel. There's enough energy in the oceans to power the world for millions of years! Imagine running your computer using a tiny drop of seawater.

6. Some fusion reactors are shaped like donuts. Scientists call this shape a "tokamak". The hot gas zooms around and around inside, getting hotter and hotter. It's like a racetrack for atoms!

7. Fusion doesn't create harmful waste like some other power plants. The main waste product is helium - the same stuff that makes balloons float! Imagine powering your home and getting party balloons as a bonus.

8. Fusion reactors need to be super cold on the outside to work. Scientists use special materials that are colder than space! It's like having the hottest thing in the universe right next to the coldest.

9. In France, scientists are building a huge fusion reactor called ITER. It's as big as 60 soccer fields! When it's done, it could show us how to power the world with fusion. It's like building a miniature sun on Earth.

10. Fusion happens when atoms smoosh together to make bigger atoms. It's the opposite of the splitting atoms in nuclear power plants. Imagine atoms playing a giant game of atomic tag, where instead of running away, they try to catch each other!

11. Some scientists think fusion reactors could one day power spaceships. These could take us to other planets much faster than regular rockets. It would be like giving the Millennium Falcon a major upgrade!

12. Fusion reactors are so hard to build, countries around the world are working together to make them work. It's like a giant, global science project where everyone shares their homework.

13. The fuel for fusion reactors is so powerful, a glass of it could power your house for hundreds of years! Imagine never having to pay an electricity bill again because you have a cupful of star-stuff in your basement.

14. Scientists use computers as powerful as thousands of laptops to design fusion reactors. These supercomputers help them figure out how to control the hot gas. It's like playing the world's most complicated video game, where the prize is unlimited energy!

15. In a fusion reactor, atoms move so fast they can circle the reactor thousands of times in one second. That's faster than the fastest roller coaster in the world! It's like atomic bumper cars at super speed.

16. Fusion reactions give off special particles called neutrons. Scientists can use these to make new materials or even clean up nuclear waste. It's like turning the power of a star into a super-cleaning machine!

17. Some people think fusion reactors could help fight climate change. They don't produce greenhouse gases like coal or oil power plants. Imagine powering the whole world without making the Earth too hot.

18. Fusion is what powers the sun and stars. By building fusion reactors, we're learning how stars work. It's like being a star detective, figuring out the secrets of the universe in our own backyard!

19. In some fusion experiments, the hot gas can reach temperatures of 150 million degrees Celsius! That's ten times hotter than the center of the sun. Imagine having something in your town that's hotter than a star!

20. Fusion reactors could one day be small enough to fit in a truck. These could power ships or even airplanes. Imagine flying around the world without ever needing to stop for fuel!

Chapter 21: Carbon capture technology

1. Imagine a giant vacuum cleaner for the air! That's kind of what carbon capture technology does. It sucks harmful carbon dioxide out of the air to help fight climate change. It's like giving Earth a big, fresh breath of air!

2. Did you know some scientists are turning captured carbon into stone? They mix it with water and pump it underground, where it turns into rock. It's like playing a magic trick on pollution, turning it into part of the Earth itself!

3. There's a company that's using captured carbon to make sneakers! They turn the carbon into a special material for shoe soles. Imagine walking around on shoes made from air pollution - it's like stepping on clouds of cleaned-up sky!

4. Some power plants have giant filters that catch carbon before it escapes into the air. It's like putting a mask on a chimney! This helps keep the air cleaner while still making electricity for our homes.

5. Scientists are creating special trees that can capture more carbon than regular trees. These super-trees grow faster and bigger, sucking up more CO2 as they grow. It's like having a forest of hungry giants eating up air pollution!

6. There's a machine that can pull carbon out of the air and turn it into fuel for cars. It's like recycling the air to power our vehicles! Imagine driving a car that runs on the same stuff that trees breathe in.

7. Some farmers are using special farming methods to trap more carbon in the soil. This helps plants grow better and keeps carbon out of the air. It's like turning farms into carbon-catching superheroes!

8. Engineers have designed buildings that can absorb carbon dioxide from the air. The buildings use special materials that soak up CO2 like a sponge. Imagine living in a house that cleans the air just by standing there!

9. Scientists are exploring ways to feed captured carbon to algae in the ocean. The algae then use it to grow, just like plants on land. It's like having a giant underwater garden that helps clean our air!

10. There's a company that turns captured carbon into baking soda. You could be baking cookies with ingredients made from cleaned-up air! It's like magic - turning pollution into something you can eat.

11. Some researchers are working on machines that mimic how trees capture carbon. These artificial trees could be placed anywhere, even in deserts! Imagine a forest of robot trees helping to clean our air.

12. Engineers have created a way to capture carbon from factory smokestacks and turn it into building materials. It's like taking the pollution and using it to build houses and schools!

13. Scientists are exploring how to use captured carbon to make stronger concrete. This could help us build sturdier buildings while also cleaning the air. It's like making Legos out of air pollution!

14. There's a plan to use giant fans to blow air through carbon-catching filters. These could be set up in places where there's lots of pollution. It's like giving the Earth its own air purifier!

15. Some researchers are looking at ways to feed captured carbon to bacteria. The bacteria could then make useful products like plastics. Imagine toys made by tiny bugs eating air pollution!

16. Scientists have found a way to turn captured carbon into carbon nanotubes - super strong and light materials. These could be used to make everything from bicycles to spaceships. It's like spinning pollution into super-strong spider webs!

17. There's a company that uses captured carbon to help grow vegetables in greenhouses. The extra CO_2 helps plants grow bigger and faster. It's like giving veggies a super-growth potion made from cleaned-up air!

18. Researchers are exploring how to use captured carbon to make medicines. Imagine taking a pill that was once harmful air pollution, now turned into something that makes you feel better!

19. Some scientists are working on ways to pump captured carbon deep into the ocean. The pressure down there turns it into a liquid that stays put. It's like giving the ocean a fizzy drink that it keeps forever!

20. Engineers have designed special paints that can absorb carbon dioxide. Painting your house with this could help clean the air around it. Imagine your bedroom walls quietly eating up pollution while you sleep!

Chapter 22: Swarm robotics

1. Imagine a team of tiny robots working together like a swarm of bees! That's what swarm robotics is all about. These little bots can do big jobs by working as a team. It's like having a bunch of robot ants that can build a skyscraper!

2. Scientists have made tiny swimming robots that move like a school of fish. They can explore underwater places that are too dangerous for people. It's like having a team of robot fish detectives searching for sunken treasure!

3. There are swarm robots that can form different shapes, like a big hand or a ladder. They connect to each other like living building blocks. Imagine having a toy set that could turn into any tool you need!

4. Some researchers are teaching swarm robots to play soccer! Each little robot is like a player on the team, passing the ball and scoring goals together. It's like watching a miniature robot World Cup!

5. Farmers are using swarm robots to help plant and care for crops. These tiny bots can check each plant and give it just what it needs. It's like having an army of robot gardeners taking care of a giant vegetable patch!

6. Scientists have made flying swarm robots that move like a flock of birds. They could be used to study the weather or help in search and rescue missions. Imagine a sky full of tiny robot birds helping to find lost hikers!

7. There are swarm robots designed to clean up oil spills in the ocean. They work together to absorb the oil and keep our seas clean. It's like having a team of robot janitors mopping up a giant water spill!

8. Researchers are working on swarm robots that can build things in space. They could construct space stations or repair satellites without astronauts having to go on dangerous spacewalks. It's like having a construction crew of mini-astronauts!

9. Some swarm robots are inspired by ants. They can carry things much heavier than themselves by working as a team. Imagine a group of robot ants helping you move your furniture around!

10. Scientists have created swarm robots that can change color like chameleons. They can blend in with their surroundings or create cool patterns. It's like having a living, moving art display made of tiny robots!

11. There are swarm robots designed to explore other planets. They could scatter across Mars or the Moon, collecting samples and taking pictures. It's like sending a robot expedition team to uncover the secrets of alien worlds!

12. Researchers are teaching swarm robots to work together in rescue missions. They could search through rubble after an earthquake, looking for survivors. Imagine a team of tiny robot heroes coming to save the day!

13. Some swarm robots are designed to work inside the human body. They could deliver medicine or help doctors see inside us better. It's like having a microscopic robot medical team exploring your insides!

14. Scientists have made swarm robots that can form bridges or stairs. This could help people cross dangerous areas safely. Imagine a group of robots turning themselves into a ladder to help you climb out of a deep hole!

15. There are swarm robots that can create 3D maps of buildings or caves. They fly or crawl around, each robot scanning a small area. It's like having a team of robot explorers drawing a map of a hidden treasure cave!

16. Researchers are working on swarm robots that can pollinate plants like bees. This could help farms if there aren't enough real bees around. Imagine a swarm of tiny robot bees helping flowers grow into yummy fruits and vegetables!

17. Some swarm robots are designed to fix other, bigger machines. They could crawl inside engines or computers and work together to make repairs. It's like having a team of mini-mechanics that can fix your car from the inside out!

18. Scientists have created swarm robots that can form words or pictures when viewed from above. They move into position like a carefully choreographed dance. Imagine a group of robots spelling out your name for your birthday!

19. There are swarm robots designed to help in factories. They can work together to move materials or assemble products. It's like having a team of tiny robot workers that never get tired and always work perfectly together!

20. Researchers are exploring how swarm robots could help in traffic management. They could become smart road markers that move to open new lanes or guide cars safely. Imagine the yellow lines on the road coming to life to help you drive better!

Chapter 23: Nanomaterials and their applications

1. Imagine a material so small you can't even see it! That's what nanomaterials are. They're tiny particles that can do amazing things. Scientists use them to make super-strong bikes, self-cleaning windows, and even socks that never smell bad! It's like having magical dust that can improve almost anything.

2. There's a special nanomaterial that can make water bounce off clothes. Spill your juice? No problem! The liquid just rolls right off. It's like wearing a force field that keeps you dry. Imagine never having to worry about stains on your favorite t-shirt again!

3. Scientists have made nanomaterials that can clean up oil spills in the ocean. These tiny particles attract oil like magnets, making it easy to scoop up. It's like having an army of microscopic janitors cleaning up the sea!

4. Some sunscreens use nanomaterials to protect your skin better. These tiny particles reflect the sun's rays, keeping you safe from sunburn. It's like wearing an invisible shield that bounces sunlight away from you!

5. Nanomaterials can make tennis balls bounce higher and last longer. The tiny particles make the rubber stronger and bouncier. Imagine playing with a tennis ball that feels like it has super powers!

6. There are nanomaterials that can make windows clean themselves! When it rains, these special particles help wash away dirt and grime. It's like having windows with built-in windshield wipers that work all by themselves.

7. Scientists use nanomaterials to make batteries that charge super fast. Your tablet could power up in minutes instead of hours! It's like giving your devices a turbo-boost so they're ready to go in no time.

8. Some bandages use nanomaterials to help cuts heal faster. These special particles can fight germs and help your skin repair itself. It's like having a tiny team of doctors right on your bandage!

9. Nanomaterials can make cars lighter and stronger. This helps them use less fuel and keep people safer in accidents. Imagine riding in a car that's as light as a feather but as strong as a tank!

10. There are nanomaterials that can change color when they detect certain chemicals. Scientists use these to make sensors that can spot pollution or spoiled food. It's like having a color-changing detector that warns you about invisible dangers!

11. Some paints use nanomaterials to resist scratches and stay shiny longer. Your bike could look brand new for years! It's like giving your stuff a super-tough invisible coat that protects it from damage.

12. Scientists are using nanomaterials to make super-thin, flexible solar panels. These could be used to make clothes that charge your phone while you wear them! Imagine having a jacket that powers your gadgets as you walk around.

13. Nanomaterials can make glass that never fogs up. No more wiping mirrors after a hot shower! It's like having magical glass that always stays clear, no matter how steamy it gets.

14. There are nanomaterials that can purify water by killing germs and removing pollution. They could help bring clean water to places that don't have it. It's like having a tiny cleanup crew that makes dirty water safe to drink!

15. Scientists use nanomaterials to make computer chips smaller and faster. This is why phones keep getting smarter and more powerful. Imagine having a computer as powerful as a giant machine, but small enough to fit in your pocket!

16. Some sports equipment uses nanomaterials to improve performance. Tennis rackets and golf clubs can be made stronger and lighter. It's like giving athletes tools with superpowers to help them play better!

17. Nanomaterials can make firefighter suits more heat-resistant. This helps keep firefighters safer when they're battling blazes. Imagine wearing a suit that protects you like a dragon's scales!

18. There are nanomaterials that can help bones heal faster after they're broken. These tiny particles help new bone grow stronger and quicker. It's like having a team of microscopic construction workers rebuilding your bones!

19. Scientists are using nanomaterials to make artificial leaves that can produce clean fuel from sunlight. It's like creating tiny factories that turn sunshine into energy we can use!

20. Some nanomaterials can make food stay fresh longer by killing bacteria. This could help reduce food waste and keep your snacks tasty for longer. Imagine having a lunchbox that keeps your sandwich fresh all day long!

Chapter 24: Tidal energy systems

1. Imagine giant underwater windmills that spin with the ocean's tides! That's what tidal energy systems are like. They use the power of moving water to make electricity. It's like having a huge water wheel in the sea, turning the ocean's movement into power for our homes!

2. Did you know the moon helps make tidal energy? Its gravity pulls on the oceans, creating tides. Tidal energy systems capture this movement. It's like the moon is playing a giant game of tug-of-war with Earth, and we're using that game to power our cities!

3. Some tidal energy systems look like big fences in the water. As the tide flows through them, they spin and make electricity. Imagine a magical fence that turns the ocean's power into lights for your house!

4. In South Korea, there's a tidal energy plant that powers over 500,000 homes! It's like a giant sea monster that eats tides and burps out electricity for an entire city.

5. Tidal energy is super reliable because tides happen every day, no matter the weather. Unlike solar panels that need sun, or wind turbines that need wind, tidal systems always have moving water. It's like having a never-ending source of power under the sea!

6. Some scientists are designing tidal kites! These underwater kites fly through strong tidal currents, pulling on cables to generate power. Imagine flying a kite deep under the ocean to make electricity!

7. In Scotland, there's a tidal turbine that looks like a giant drill. It spins in the fast-moving water between islands. It's so powerful it can supply electricity to 2,000 homes! That's like powering a whole neighborhood with one underwater pinwheel.

8. Tidal energy systems can also protect coasts from flooding. Some act like barriers, stopping storm surges while also making electricity. It's like having a superhero that guards the coast and powers your TV at the same time!

9. Did you know some tidal energy systems look like giant snakes? They float on the water's surface and move up and down with the waves, generating power. Imagine a friendly sea serpent that makes electricity every time it wiggles!

10. In France, there's a tidal power plant that's been running since 1966! It's like a grandpa of tidal energy, still going strong after all these years. It shows how tidal power can last for a long, long time.

11. Some tidal systems use special buoys that bob up and down with the waves. As they move, they pump seawater through turbines to make electricity. It's like having a bunch of giant rubber ducks in the ocean, all working together to power your home!

12. Tidal energy is clean and doesn't pollute the air like burning coal or oil. It's a way to make electricity without making the Earth too hot. Imagine powering your video games with the ocean's waves instead of smoky factories!

13. In Nova Scotia, there's a tidal turbine that fish can safely swim through. Scientists made sure it wouldn't hurt ocean life. It's like building an underwater playground that fish can enjoy while it makes power for us!

14. Some people are working on tidal systems that look like underwater carousels. They spin slowly with the tides, generating electricity as they turn. Imagine a merry-go-round for fish that also powers your lights!

15. Tidal energy can be used on rivers too! In New York, there are turbines in the East River that make power from the strong currents. It's like putting tiny waterwheels in a river to catch its energy as it flows by.

16. Scientists are designing tidal systems that look like giant leaves. They sway back and forth with the water's movement, creating energy. Imagine an underwater forest of metal leaves, all working together to make clean electricity!

17. In Japan, they're testing a system that uses temperature differences between shallow and deep water to make energy. It's like taking advantage of the ocean's hot and cold spots to power our homes!

18. Some tidal energy systems can pump water into reservoirs when there's extra power. Later, this water can flow back down to make more electricity when it's needed. It's like filling a giant bathtub with the ocean and then letting it drain to power a city!

19. In Canada, they're working on a tidal system that floats like a raft. It moves up and down with the tides, using this motion to generate power. Imagine riding a raft that makes electricity every time a wave passes by!

20. Tidal energy systems can help small islands that are far from other power sources. They can provide a steady supply of electricity without needing to bring in fuel. It's like giving these islands their own personal power plants, powered by the sea around them!

Conclusion

Wow! What an incredible journey through the world of cutting-edge engineering we've had together! We've explored amazing inventions that sound like they're straight out of science fiction, but they're actually real or in the works right now. From the tiniest nanomaterials to enormous fusion reactors, from the depths of the ocean to the edges of space, we've seen how engineers are dreaming up solutions to some of the world's biggest challenges.

Remember, every great invention starts with a curious mind and a big imagination — just like yours! The next time you look at a problem, think about how you might solve it in a creative way. Could you use swarm robotics to clean up your room faster? Or maybe design a vertical farm to grow fresh veggies in your backyard?

The future is full of possibilities, and who knows? You might be the one to invent the next amazing thing we can't live without. So keep asking questions, stay curious, and never stop imagining what could be. The world needs your bright ideas and creative thinking!

Now, go out there and look at the world with new eyes. There's engineering magic all around us, just waiting to be discovered. Who's ready to be the next great inventor?

www.ingramcontent.com/pod-product-compliance
Lightning Source LLC
Chambersburg PA
CBHW072051230526
45479CB00010B/675